YOUR KNOWLEDGE HAS VALUE

Bibliographic information published by the German National Library:

The German National Library lists this publication in the National Bibliography; detailed bibliographic data are available on the Internet at http://dnb.dnb.de .

Imprint:

Copyright © 2015 GRIN Verlag, Open Publishing GmbH
Print and binding: Books on Demand GmbH, Norderstedt Germany
ISBN: 9783668249653

This book at GRIN:

http://www.grin.com/en/e-book/335001/an-experimental-investigation-for-the-coolant-temperature-effect-on-the

Mina Abaskharon, Fawzy M. H. Ezzat, Ali M. Abd-El-Tawwab, Mohamed R. El-Sharkawy

An Experimental Investigation for the Coolant Temperature Effect on the Exhaust Emissions for a Spark Ignition Engine Fuelled with Gasoline and CNG

GRIN Publishing

GRIN - Your knowledge has value

Since its foundation in 1998, GRIN has specialized in publishing academic texts by students, college teachers and other academics as e-book and printed book. The website www.grin.com is an ideal platform for presenting term papers, final papers, scientific essays, dissertations and specialist books.

Visit us on the internet:

http://www.grin.com/

http://www.facebook.com/grincom

http://www.twitter.com/grin_com

International Journal of Scientific & Engineering Research, Volume 6, Issue 10, October-2015
ISSN 2229-5518

An experimental investigation for the coolant temperature effect on the exhaust emissions for a spark ignition engine fuelled with gasoline and CNG

Mina B. R. Abaskharon, Fawzy M. H. Ezzat, Ali M. Abd-El-Tawwab, Mohamed R. El-Sharkawy

Abstract— In the present work a comparative assessment has been made for the exhaust emissions of a spark ignition engine fuelled with gasoline and CNG. The engine under test was operated separately by gasoline or CNG using a conversion switch. The produced hydrocarbon (HC), carbon monoxide (CO) and carbon dioxide (CO_2) of both fuels were measured at coolant temperature of 80℃, 90℃ and 100℃. Tests have been conducted at full and half load operating conditions with a speed range from 1000:5000 rpm. The results showed that reducing the coolant temperature from 100℃ to 80℃ increased the produced hydrocarbon and carbon dioxide and reduced the carbon monoxide for both fuels at full and half load conditions. Furthermore, the CNG produced less HC, CO and CO_2 than the gasoline at full and half load operating conditions.

Index Terms— CNG, Gasoline, Emissions, Hydrocarbon, Carbon monoxide, Carbon dioxide, Coolant temperature, Combustion.

———————————— ◆ ————————————

1 INTRODUCTION

Natural gas is a promising alternative fuel to meet strict engine emission regulations in many countries. Compressed natural gas (CNG) has long been used in stationary engines, but the application of CNG as a transport engines fuel has been considerably advanced over the last decade by the development of lightweight high-pressure storage cylinders. Engine conversion technology is well established and suitable conversion equipment is readily available [1]. M. U. Aslam and co-workers reported that on average, the retrofitted CNG engine reduced the CO by around 80%, CO_2 by 20% and HC by 50% and increases the NO_X emissions by around 33% in comparison with gasoline engine [2]. M .I. Jahirul et al [3], studied the exhaust gases released by cars in Malaysia using liquid fuels and natural gas between 2006 and 2020. It was found that CNG has much lower emission levels than gasoline or diesel fuels. The study revealed that from the use of CNG as motor fuel, the air pollution reduction can be as high as 90% from passenger cars in Malaysia. Shamekhi et al [4],

- Mina Badrat Rezq Abaskharon is a Demonstrator in Automotive and Tractors Engineering Department, Faculty of Engineering, Minia University, Minia, Egypt, and pursuing master degree program in internal combustion engine field. PH: +201226545808. E-mail: minabadrat@gmail.com

- Fawzy Mohammed Hashem Ezzat is a Professor of Automotive and Tractors Engineering Department, Faculty of Engineering, Minia University, Minia, Egypt.

- Ali Mahmoud Abd EL-Tawwab Ali is a Professor of Automotive and Tractors Engineering Department, Faculty of Engineering, Minia University, Minia, Egypt.

- Mohamed Rashad Mohamed El- Sharkawy is a Lecturer of Automotive and Tractors Engineering Department, Faculty of Engineering, Minia University, Minia, Egypt.

investigated that using the CNG as a fuel reduced the CO and CO . CO emissions decreased between 58% and 89% and the CO between 0% and 11%. The HC emissions demonstrate reduction between 37% and 58%. The NO_X emissions were the only ones that increased with the CNG. Exhaust emissions for a retrofitted CNG engine were measured by How Heoy Geok et al [5], it was found that emission of pollutant gaseous from automotive engine was significantly reduced by the use of CNG with 40-87% reduction of unburned hydrocarbons. The reduction of CO and CO_2 emissions were 20-98% and 8-20% respectively using CNG. M.I. Jahirul et al [6], experimentally studied the performance and exhaust emission on a gasoline and CNG fuelled retrofitted spark ignition engine. Lower HC and CO were produced by using the CNG throughout the speed range. Higher NO_X emissions were the main emission concerned for CNG as automotive fuel. 41% and 38% higher NO_X emissions have been recorded at 50% and 80% throttle position respectively, compared to that of gasoline.   A. Rehman et al [7] investigated the influence of coolant temperature on the Performance of a four stroke spark ignition engine with a dual circuit cooling system. It was observed that the higher coolant temperature around the cylinder block decreased the amount of HC due to the reduced wall quenching effect. Results revealed about 12-15% reductions in HC emissions. Santhosh Thomas et al [8], investigated the effect of coolant temperature on the performance and emissions of naturally aspirated gasoline engine. Lower coolant temperature increased the CO_2 percentage, and decreased the CO percentage emission. But this advantage was offset by the increase in the total hydrocarbon emissions. Lowering the coolant temperature has a slight advantage on the NO_X emission. Dashti Mehrnoosh et

International Journal of Scientific & Engineering Research, Volume 6, Issue 10, October-2015
ISSN 2229-5518

al [9], carried out a thermodynamic cycle simulation of a conventional four-stroke SI engine using gasoline and CNG fuels.it was found that the CO_2, CO and concentration of HC were decreased considerably by about 33%, 60% and 53%, respectively, while NO_x concentration increased by 50%. Ajay K. Singh et al [10], experimentally studied the effect of the engine coolant temperature on exhaust emission of four stroke spark ignition multi cylinder engine. The study confirmed that exhaust emission was a dependent parameter on the engine coolant temperature. Raising the temperature of coolant in the engine block can produce significant improvements in the engine performance with a corresponding reduction in Hydrocarbon emission. Also lowering the coolant temperature in the cylinder head increased the knock limit of the engine with a corresponding reduction in the levels of NO_x in exhaust emissions. Saad Aljamali et al [11], investigated a comparison of performance and emission of a gasoline engine fuelled by gasoline and CNG. The emission of NO and NO_x of CNG was lower at the low and high engine speeds. The emissions of CO_2 and CO were found less for the CNG compared to gasoline. In this study, a comparative evaluation for the emissions of gasoline and CNG fueled engine with respect to the coolant temperature has been performed.

2 EXPERIMENTAL SETUP

The layout of the experimental setup has been shown in figure (1). The engine under test was spark ignition engine and

was converted to a bi-fuel (gasoline and CNG) engine at automotive and tractors engineering department laboratory, faculty of engineering, Minia University, Egypt. The main specifications of the test engine were listed in table [1].

Table [1]. Engine specifications

Engine model	Hyundai (G4EH)
Engine type	Gasoline
Displacement (cm³)	1341
Number of cylinder	4
Compression ratio	9.5
Bore (cm)	7.15
Stroke (cm)	8.35
Max.power (Kw/rpm)	61.78/5500
Max.torque (Nm/rpm)	116.7/3000
Cycle	Four
Fuel system	Multipoint injection

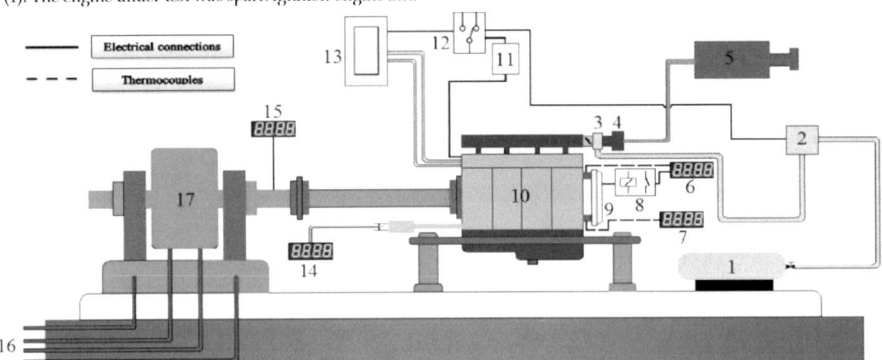

Fig.1: Engine test bed

1: CNG cylinder	2: CNG regulator
3: CNG mixer	4: Air filter
5: Intake air surge chamber	6: Temperature measurement & controller at the coolant outlet port
7: Temperature measurement at the coolant inlet port	8: Relay for the fan operation
9: Radiator & fan	10: SI engine
11: Spark timing advance processor	12: Conversion switch
13: Gasoline vessel with fuel pump	14: Exhaust emissions analyzer
15: Digital tachometer	16: Water paths for hydraulic dynamometer
17: Hydraulic dynamometer	

The used CNG kit was the conventional (mixer) kit. The engine was work separately by gasoline or CNG using a conversion switch. A hydraulic dynamometer (HPA engine dynamometer 203) was connected to the engine to measure the engine load. The specifications of the used gasoline and CNG were listed in tables (2, 3). Emissions parameter such as hydrocarbon (HC), Carbon monoxide (CO) and Carbon dioxide (CO_2) were measured using KANE AUTO 4-2 exhaust gas analyzer.

Table [2]. Gasoline specification [12].

Properties	Limits
Research octane number (MIN)	92
Lead content (MAX)	.013 (g/l)
Existent gum content (MAX)	3 (mg/100 ml)
Total Sulphur (MAX)	.05 (% Mass)
Gross heating value	47589 kj/kg
Net heating value	44310 kj/kg

Table [3]. CNG specification [13].

Component	Symbol	Value	Unit
Nitrogen	N_2	.6120	
Methane	CH_4	90.038	
Ethane	C_2H_6	5.1380	
Propane	C_3H_8	.6260	
Iso butane	C_4H_{10}	.189	Mole (%)
Iso pentane	C_5H_{12}	.068	
Hexanes	C_6H_{14}	.0650	
Carbon dioxide	CO_2	3.2640	
Gross heating value	GHV	1028.2	(Btu/ft³)
Net heating value	NHV	931.58	(Btu/ft³)

Engine coolant temperature was measured at the inlet and outlet port between the engine and the radiator. The outlet Coolant temperature was controlled using a PID temperature controller (REX-C100) with K type thermocouple. The PID controller activated the radiator fan to maintain the coolant temperature at a certain limit, figure (2).

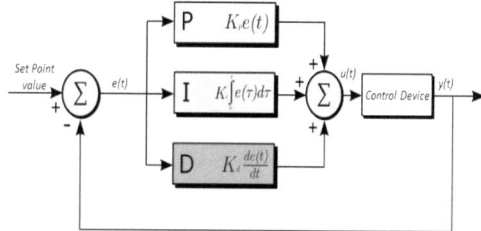

Figure (2): PID temperature controller diagram

A compression test has been made for the engine before doing the experiments to determine the engine case and the results were as follow

Table [4]. Compression test results.

Cylinder number	Pressure
Cylinder 1	195 psi
Cylinder 2	188 psi
Cylinder 3	185 psi
Cylinder 4	190 psi

3 TEST METHDOLOGY

To determine the optimum condition for the emissions of the bi-fuel engines three cases were considered in this study. The first case was the coolant temperature effect on the exhaust emissions for the gasoline operation. The second case was the coolant temperature effect on the exhaust emissions for the CNG operation. The third case was an exhaust emissions investigation for the both fuels at constant coolant temperature. The three cases were measured at the full and half load conditions from 1000:5000 rpm. Speeds over 5000 rpm were avoided for the safety concern

4 RESULTS AND DISSCUSSION

4.1 Gasoline operation emissions at different coolant temperature.

During this test, engine emissions were measured repeatedly at coolant temperature of 80°C, 90°C and 100°C at the engine outlet port. All the temperature accuracies were within ±2°C. Figures 3 to 5 show the effect of the coolant temperature on the exhaust emissions at the full load condition. It has been observed that reducing the coolant temperature from 100°C to 80°C increased the amount of hydrocarbons emissions. Cylinder wall temperature of an engine has a direct relation with the coolant temperature.

International Journal of Scientific & Engineering Research, Volume 6, Issue 10, October-2015
ISSN 2229-5518

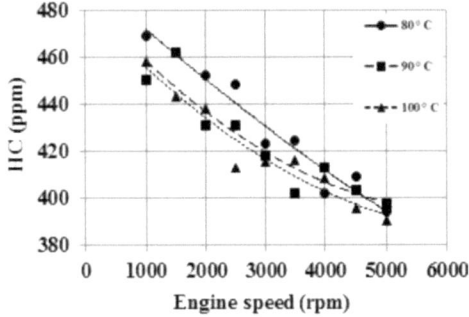

Fig.3: Hydrocarbon emissions versus engine speed at full load with gasoline

So the lower coolant temperature reduced the cylinder wall temperature and increased the flame quenching effect which increased the amount of the unburned hydrocarbon [10]. Also decreasing the coolant temperature from 100°C to 80°C decreased the carbon monoxides and increasing the carbon dioxide in the exhaust emissions. This gave an indication of the more complete combustion with the 80 °C coolant temperature. Coolant temperature affected on the temperature of all the engine parts, especially on the intake manifold because the engine under test has a coolant path in the throttle body. The lower coolant temperature reduced the intake manifold temperature and increased the volumetric efficiency [8], this ensured the profusely presence of the oxygen and enhanced the oxidation of the carbon monoxide to the carbon dioxide. Figures (6 to 8) show the effect of the coolant temperature on the exhaust emissions at half load condition with gasoline operation

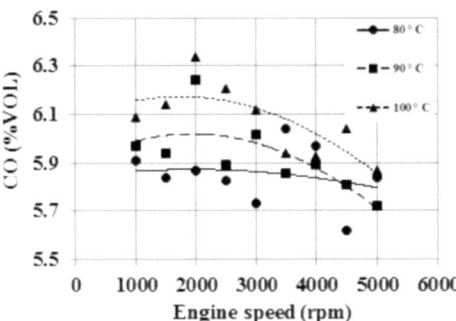

Fig.4: carbon monoxide emissions versus engine speed at full load with gasoline

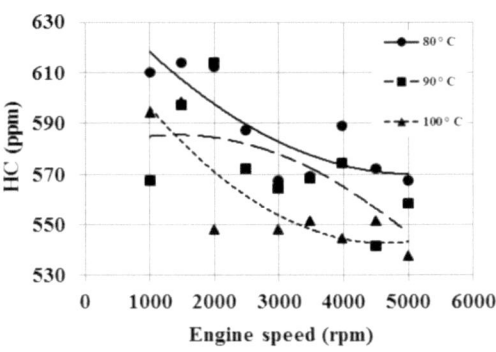

Fig.6: Hydrocarbon emissions versus engine speed at half load with gasoline

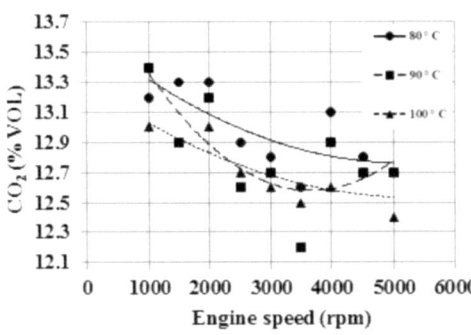

Fig.5: carbon dioxide emissions versus engine speed at full load with gasoline

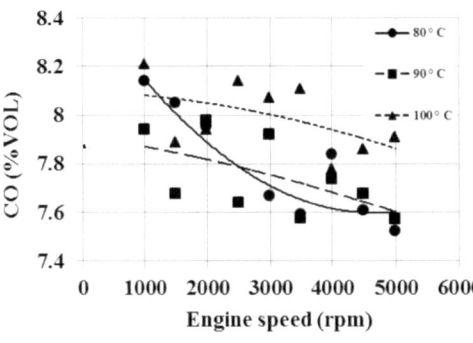

Fig.7: carbon monoxide emissions versus engine speed at half load with gasoline

International Journal of Scientific & Engineering Research, Volume 6, Issue 10, October-2015
ISSN 2229-5518

It has been detected that reducing the coolant temperature increased the hydrocarbon in the exhaust emissions because of increasing of the quench layer at the cylinders walls. Lower coolant temperature reduced the carbon monoxide and increased the carbon dioxides because of the higher volumetric efficiency at lower coolant temperature.

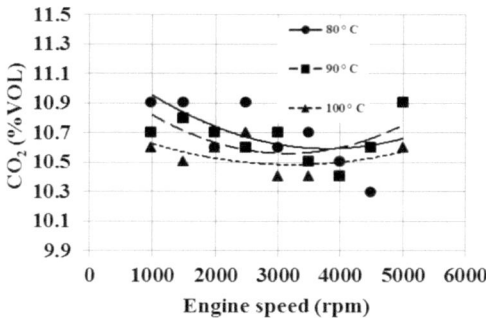

Fig.8: carbon dioxide emissions versus engine speed at half load with gasoline

The half load operation condition produced more hydrocarbon and carbon monoxide than the full load condition. The carbon dioxide which performed a sign of the complete combustion has been reduced at half load condition. So the combustion process tended to be less efficient at half load condition because of increasing of the residual gases, also the reduction of the turbulence and swirl which reduced the mixture homogeneity

4.2 CNG operation emissions at different coolant temperature.

During this test engine emissions were measured repeatedly at coolant temperature of 80°C, 90°C and 100°C at the engine outlet port. All the measurements were within ±2°C. Figures (9 to 11) show the effect of the coolant temperature on the exhaust emissions at the full load condition.

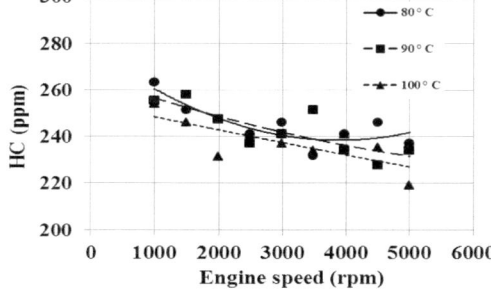

Fig.9: Hydrocarbon emissions versus engine speed at full load with CNG

It has been detected that decreasing the coolant temperature from 100°C to 80°C increased the hydrocarbon in the exhaust emissions because of increasing of flame quenching phenomena at the cylinders walls at the lower coolant temperature.

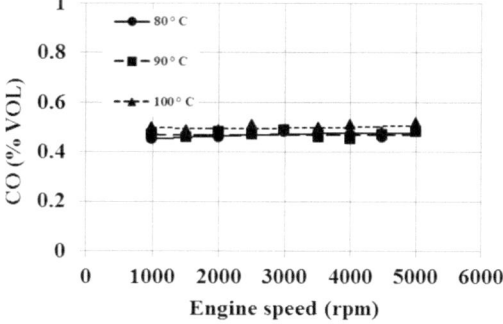

Fig.10: carbon monoxide emissions versus engine speed at full load with CNG

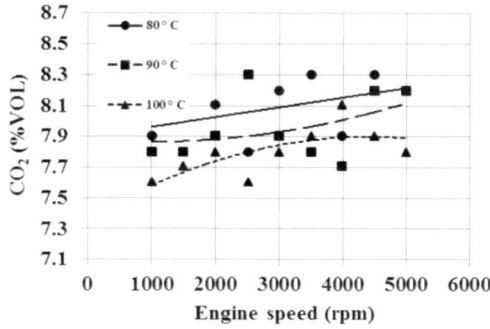

Fig.11: carbon dioxide emissions versus engine speed at full load with CNG

Reducing the coolant temperature reduced the carbon monoxide and increased the carbon dioxide in the exhaust emissions. This due to the increasing of the volumetric efficiency which secured the needed oxygen for the carbon monoxide oxidation. The design of the CNG reducer also contributed in the variation of CO and CO_2 with the coolant temperature. As the reducer has a path from the engine coolant circuit to prevent freezing due to the pressure reduction, the outlet gas properties were more sensitive to the variation in the coolant temperature especially because the high molecular distance in the gases. Figures (12 to 14) show the effect of the coolant temperature on the exhaust emissions at the half load.

International Journal of Scientific & Engineering Research, Volume 6, Issue 10, October-2015
ISSN 2229-5518

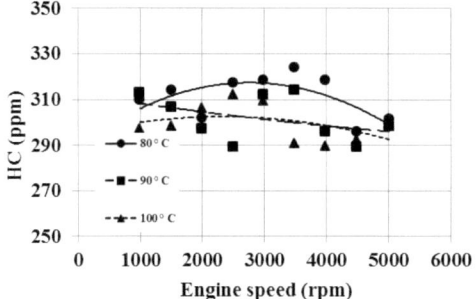

Fig.12: Hydrocarbon emissions versus engine speed at half load with CNG

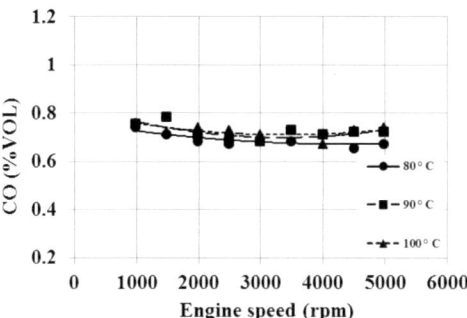

Fig.13: carbon monoxide emissions versus engine speed at half load with CNG

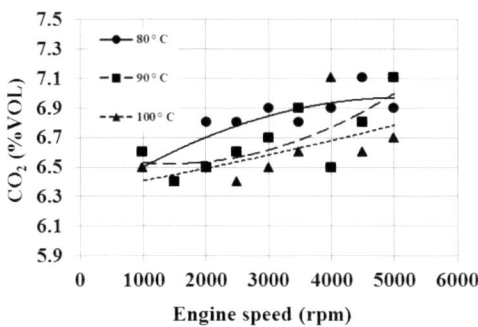

Fig.14: carbon dioxide emissions versus engine speed at half load with CNG

It has been observed that reducing the coolant temperature from 100°C to 80°C increased the hydrocarbon in the exhaust emissions because the increasing of the quench layer at the cylinders walls. The higher volumetric efficiency at the lower coolant temperature decreased the carbon monoxides and increased the carbon dioxides. Generally, decreasing the load to the half has a negative effect on the exhaust emissions because of increasing the residual gas at half load.

Figures (15, 16) show the effect of the coolant temperature on the exhaust emissions with CNG at the maximum power (43.91 Kw), 5000 rpm.

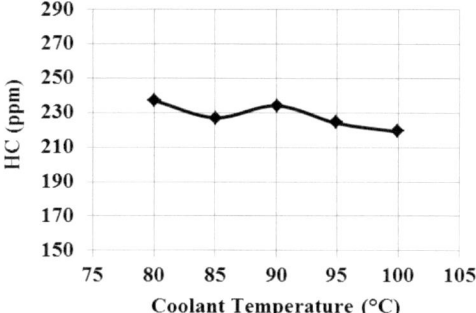

Fig.15: hydrocarbon concentration versus coolant temperature at 5000 rpm

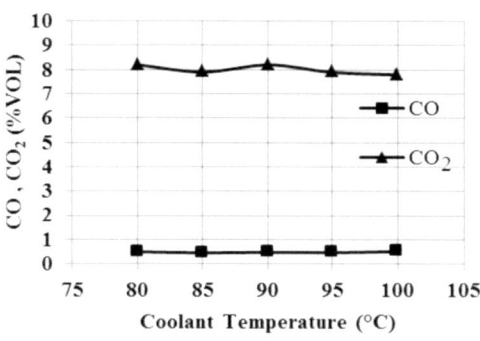

Fig.16: CO and CO_2 concentration versus coolant temperature at 5000 rpm

A little significant effects were noticed for HC and CO_2 with the coolant temperature change, for example HC changed from 237 ppm at 80°C to 219 ppm at 100°C. CO2 showed a non-substantial changes as temperature varied from 80°C to 100°C. CO showed no variations in values.

International Journal of Scientific & Engineering Research, Volume 6, Issue 10, October-2015
ISSN 2229-5518

4.3 CNG emissions compared to gasoline emissions

This section presents a comparative evaluation for the exhaust emissions of the both fuels at 90°C coolant temperature at the engine outlet port. Figures (17 to 19) indicate the effect of using CNG on the exhaust emissions such as hydrocarbon, carbon monoxide and carbon dioxide in comparison with gasoline at the full load condition

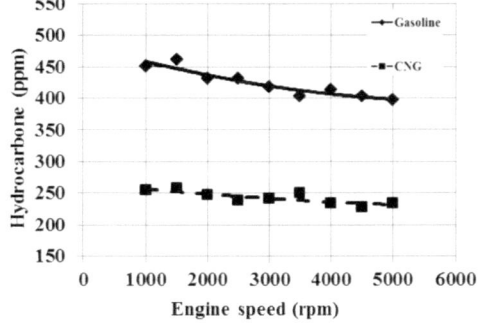

Fig.17: Hydrocarbon versus engine speed at full load, 90°C coolant temperature

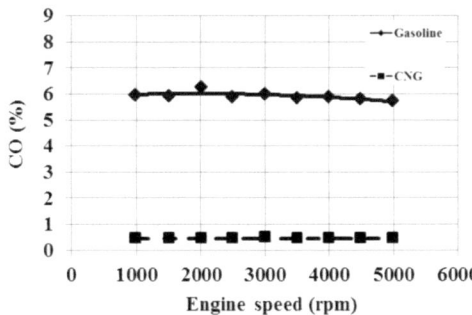

Fig.18: carbon monoxide versus engine speed   at full load, 90°C coolant temperature

It has been observed that CNG produced lower hydrocarbon, carbon monoxide and carbon dioxide emissions throughout the speed range as compared to gasoline. This is mainly due to that gaseous fuels such as CNG can form much better homogenous mixture and tends to more complete combustion than liquid fuels which require time for complete atomization and vaporization [6]. CNG also has a better air mixing characteristic because its molecular weight is less than that for gasoline [6].

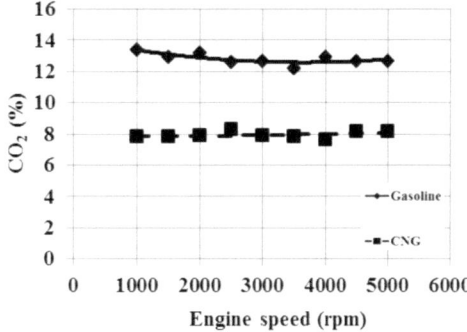

Fig.19: carbon dioxide versus engine speed at full load, 90°C coolant temperature

Hydrocarbon concentration was higher at lower engine speeds and decreased as the engine speed increased because the turbulence and swirl intensity is a strong function of engine speed. Increasing the turbulence and swirl enhanced the rate of evaporation, mixing and combustion [14]. Slightly reduction of CO with the increasing in engine speed has been noticed for the same reason. Also the different in hydrogen to carbon ratio for both fuels (4 for CNG and 2.03 for gasoline) caused that gasoline produced more CO_2 than CNG all over the speed range [15]. The same reason contributed toward the higher hydrocarbon and CO for the gasoline operation. Figures (20 to 22) show the exhaust emissions for both fuels at the half load condition.

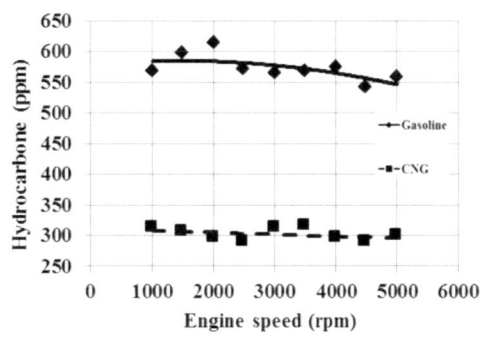

Fig.20: Hydrocarbon versus engine speed at half load, 90°C coolant temperature

CNG produced less hydrocarbon and carbon monoxide at the half load for the same previous reasons. Also less carbon dioxide was produced with CNG at half load.

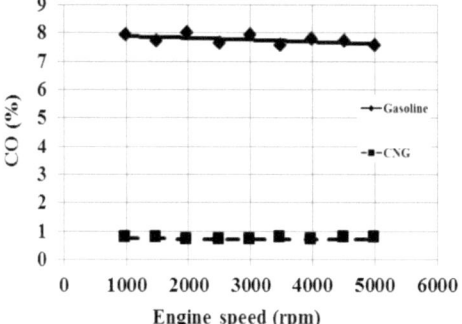

Fig.21: carbon monoxide versus engine speed at half load, 90°C coolant temperature

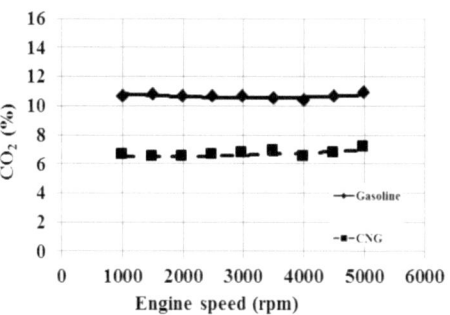

Fig.22: carbon dioxide versus engine speed at half load, 90°C coolant temperature

It has been observed that for both fuels the half load condition has more hydrocarbon, carbon monoxide and carbon dioxide than the full load condition. This because the half load was obtained at different throttle positions less than the wide open throttle which reduced the initial pressure in the cycle, turbulent and air-fuel mixing and increased the amount of residual gases, causing poor combustion at part load. [14]

5. Conclusions

1- For the CNG full load operation, reducing the coolant temperature from 100°C to 80°C increases the HC by 1.66-5.39% and CO2 by 2.96-4.77% and reduces the CO by 5.88-10% throughout the speed range.

2- For the CNG half load operation, reducing the coolant temperature from 100°C to 80°C increases the HC by 1.96-5.03% and CO2 by 1.23-4.05% and reduces the CO by 5.55-5.26% throughout the speed range.

3- At the CNG full load operation with 5000 rpm, the lower HC and CO2 was at 100 °C

4- At full load condition with 90°C coolant temperature, using the CNG reduces the HC by 42.5-44.56%, CO2 by 37-39.84% and CO by 91.58-91.66% in comparison with the gasoline operation throughout the speed range.

5- At half load condition with 90°C coolant temperature, using the CNG reduces the HC by 45.97-47.86%, CO2 by 35.77-39.44% and CO by 91.2-91.22% in comparison with the gasoline operation throughout the speed range.

References

[1] Semin, Rosli Abu Bakar, "A Technical review of compressed natural gas as an alternative fuel for internal combustion engines", American J. of Engineering and Applied Sciences, ISSN 1941-7020,pp. 302, (2008).

[2] M. U. Aslam, H. H. Masjuki, M. A. Kalam , M. A. Amalina,"A comparative evaluation of the performance and emissions of a retrofitted spark ignition car engine", Journal of Energy & Environment 4, pp. 109, (2005).

[3] M .I. Jahirul, R. Saidur, M. Hasanuzzaman, H.H. Masjuki, M.A.Kalam,"A comparison of the air pollution of gasoline and CNG driven car for malaysia", International Journal of Mechanical and Materials Engineering, Vol. 2, No. 2, pp.130,137 (2007).

[4] Shamekhi, Amir Hossein, Khatibzadeh, Nima, "A comprehensive comparative investigation of compressed natural gas as an alternative fuel in a bi-fuel spark ignition engine" Iran. J. Chem. Chem. Eng., Vol. 27, No.1, pp.82. (2008)

[5] How Heoy Geok, Taib Iskandar Mohamad, Shahrir Abdullah, Yusoff Ali, Azhari Shamsudeen, Elvis Adril, "Experimental investigation of performance and emission of a sequential port injection natural gas engine", European Journal of Scientific Research, Vol.30, No.2, ISSN1450-216X, pp. 213, (2009).

[6] M.I. Jahirul, H.H. Masjuki, R. Saidur, M.A. Kalam, M.H. Jayed, M.A. Wazed, "Comparative engine performance and emission analysis of CNG and gasoline in a retrofitted car engine",Applied Thermal Engineering 30,pp. 2225, (2010)

[7] A. Rehman, R.M.Sarviya, Savita Dixit, Rajesh Kumar Pandey, "the influence of coolant temperature on the performance of a four stroke spark ignition engine employing a dual circuit cooling system", CIGR journal, Vol.12,No.1,pp.9, (2010)

[8] Santhosh Thomas, Agam Saroop, Ranjeet Rajak, Saravanan Muthiah, "Investigation on the effect of coolant temperature on the performance and emissions of naturally aspirated gasoline engine",SIAT, India, pp.5, (2011)

[9] Dashti Mehrnoosh, Hamidi Ali Asghar and Mozafari Ali Asghar, "Thermodynamic model for prediction of performance and emission characteristics of SI engine fuelled by gasoline and natural gas with experimental verification", Journal of Mechanical Science and Technology, pp.2213, 2223. (2012)

International Journal of Scientific & Engineering Research, Volume 6, Issue 10, October-2015
ISSN 2229-5518

[10] Ajay K. Singh, Dr A. Rehman," An experimental investigation of engine coolant temperature on exhaust emission of four stroke spark ignition multi cylinder engine", pp.224, (2013)

[11] Saad Aljamali, Wan Mohd Faizal Wan Mahmood, Shahrir Abdullah, Yusoof Ali, "Comparison of performance and emissions of a gasoline engine fuelled by gasoline and CNG under various throttle positions",Journal of Applied Science, pp.390, (2014)

[12] The Egyptian general petroleum corporation, quality and control department, 2015.

[13] Egyptian natural gas company (GASCO), 2015.

[14] Willard, W. Pulkrabek, "Engineering Fundamentals of the Internal Combustion Engine", University of Wisconsin, Platteville, pp. 207,280, (2004).

[15] John G. Ingersoll, "Natural gas vehicles", pp.29, 34, (1995)

YOUR KNOWLEDGE HAS VALUE

- We will publish your bachelor's and
  master's thesis, essays and papers

- Your own eBook and book -
  sold worldwide in all relevant shops

- Earn money with each sale

Upload your text at www.GRIN.com
and publish for free